MOBILE APP WIREFRAME

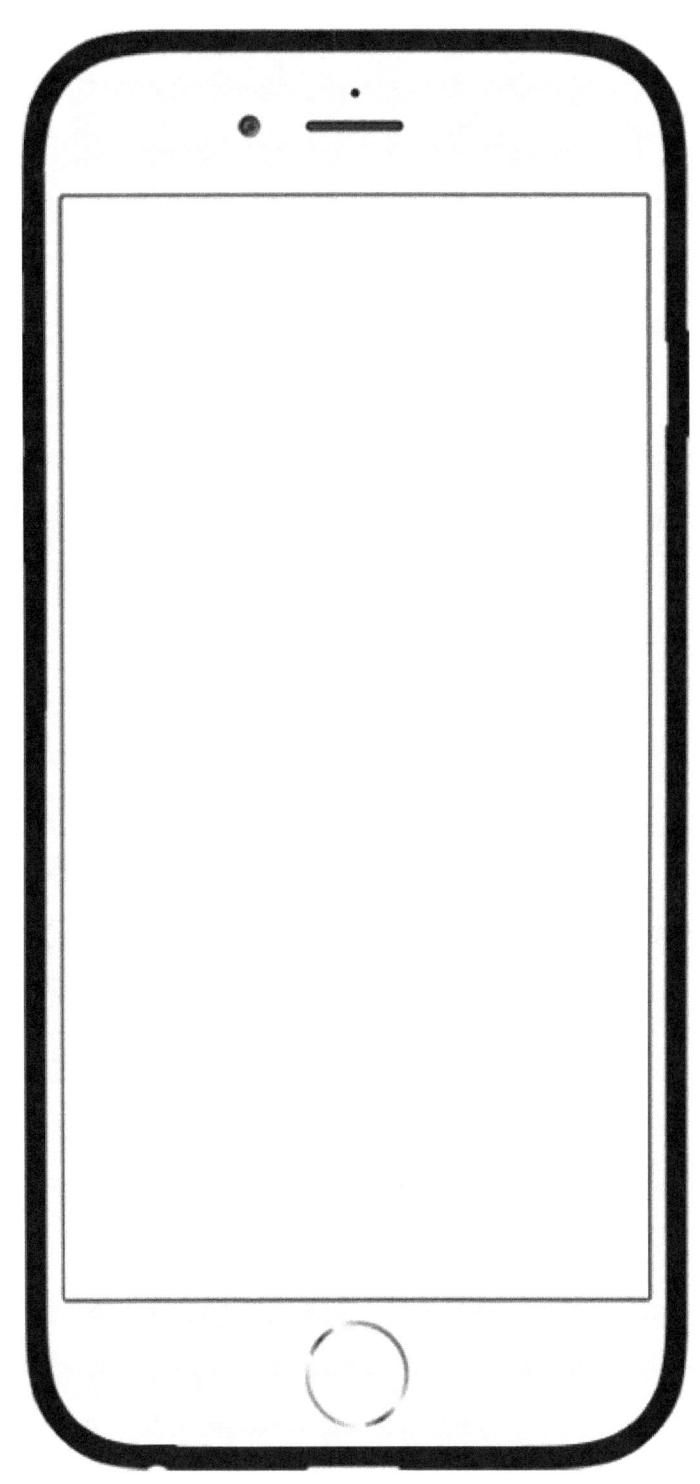

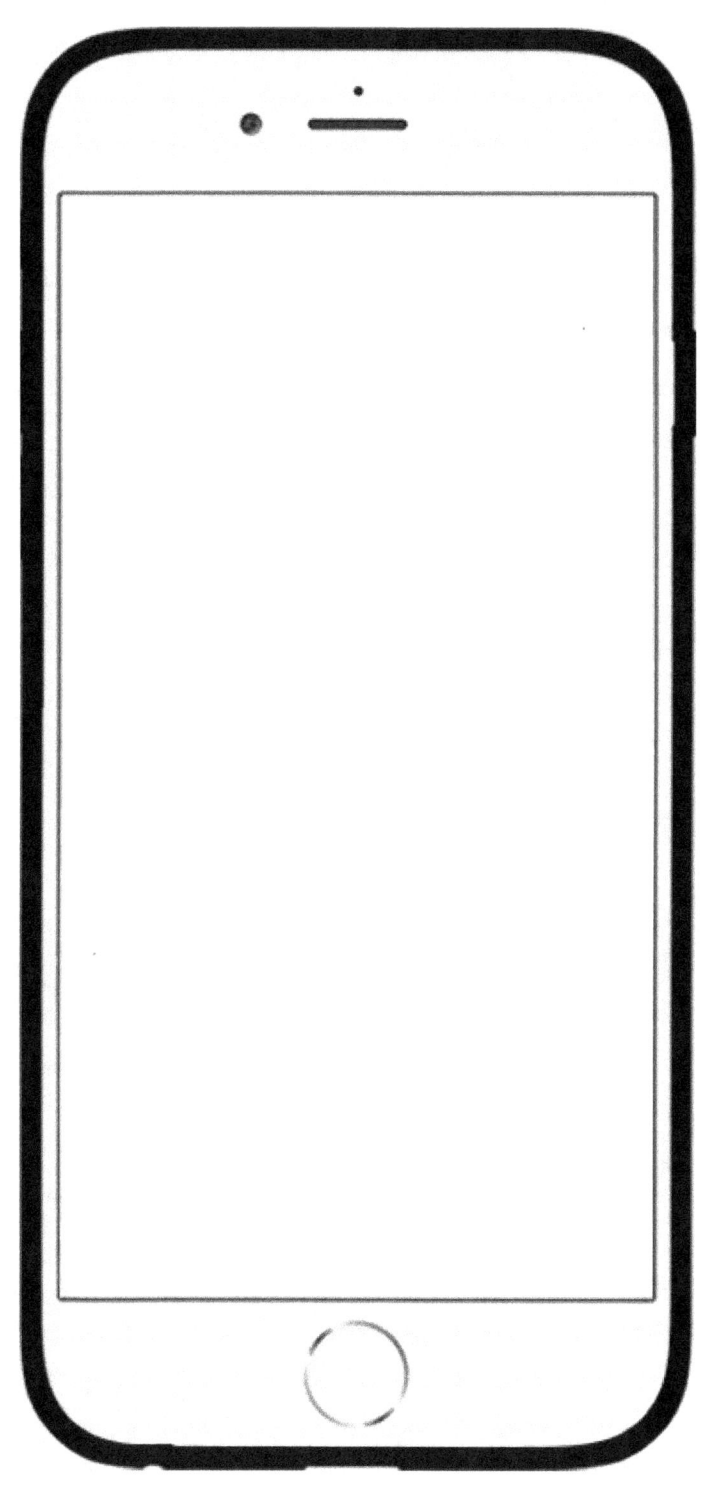

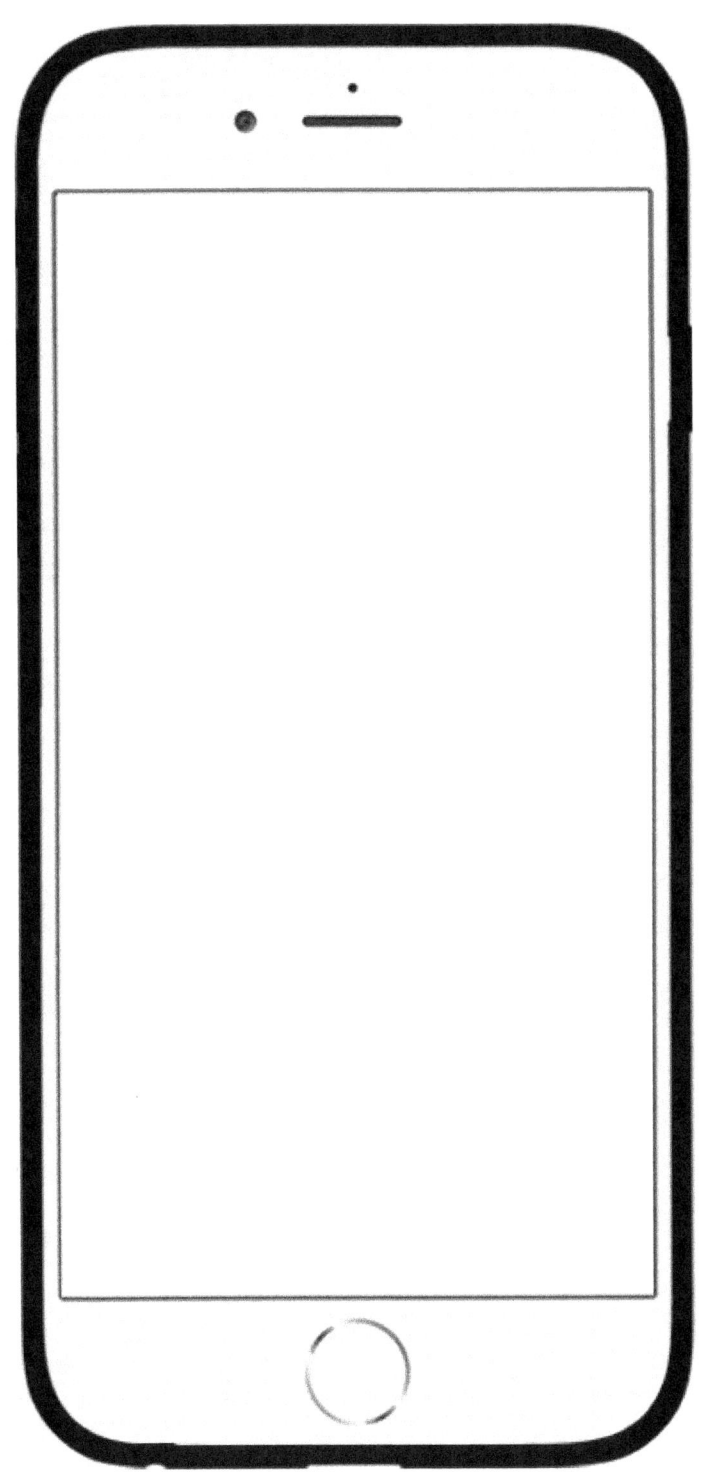

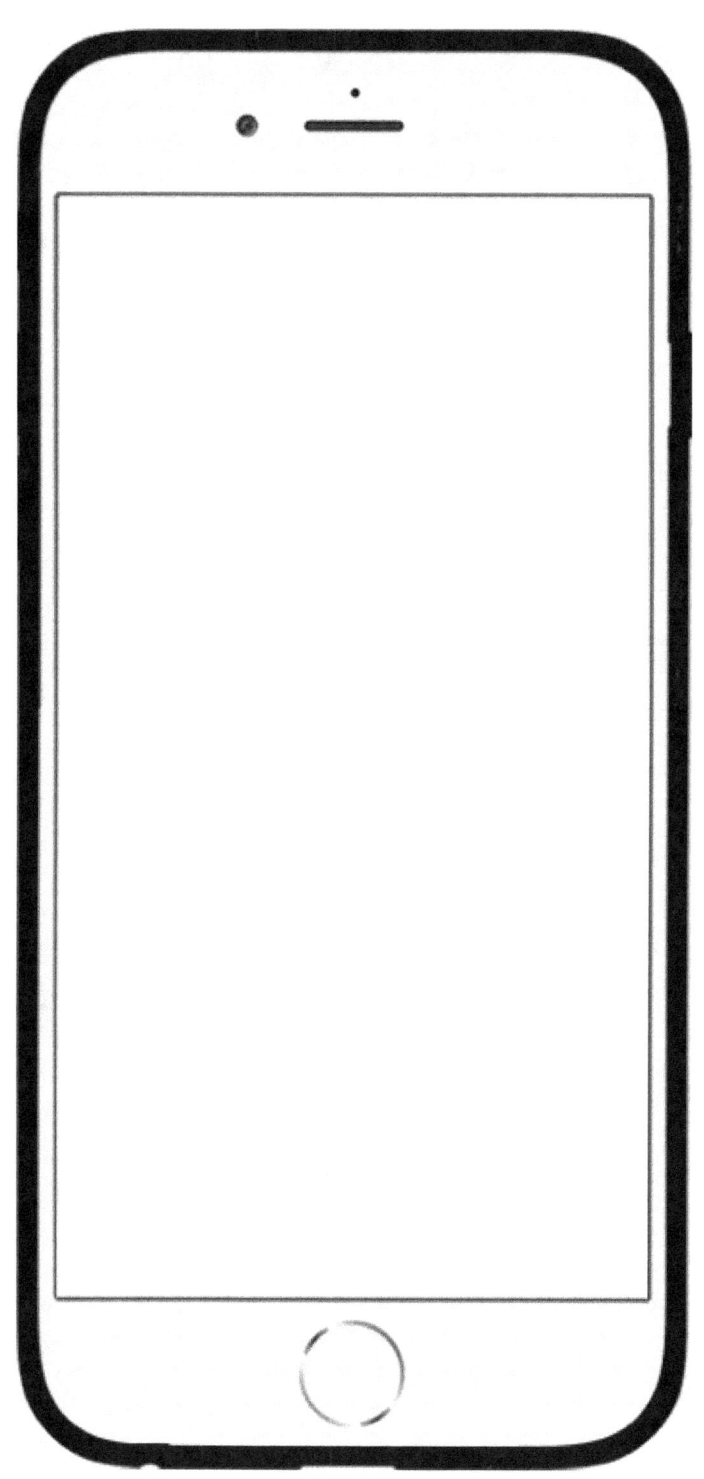

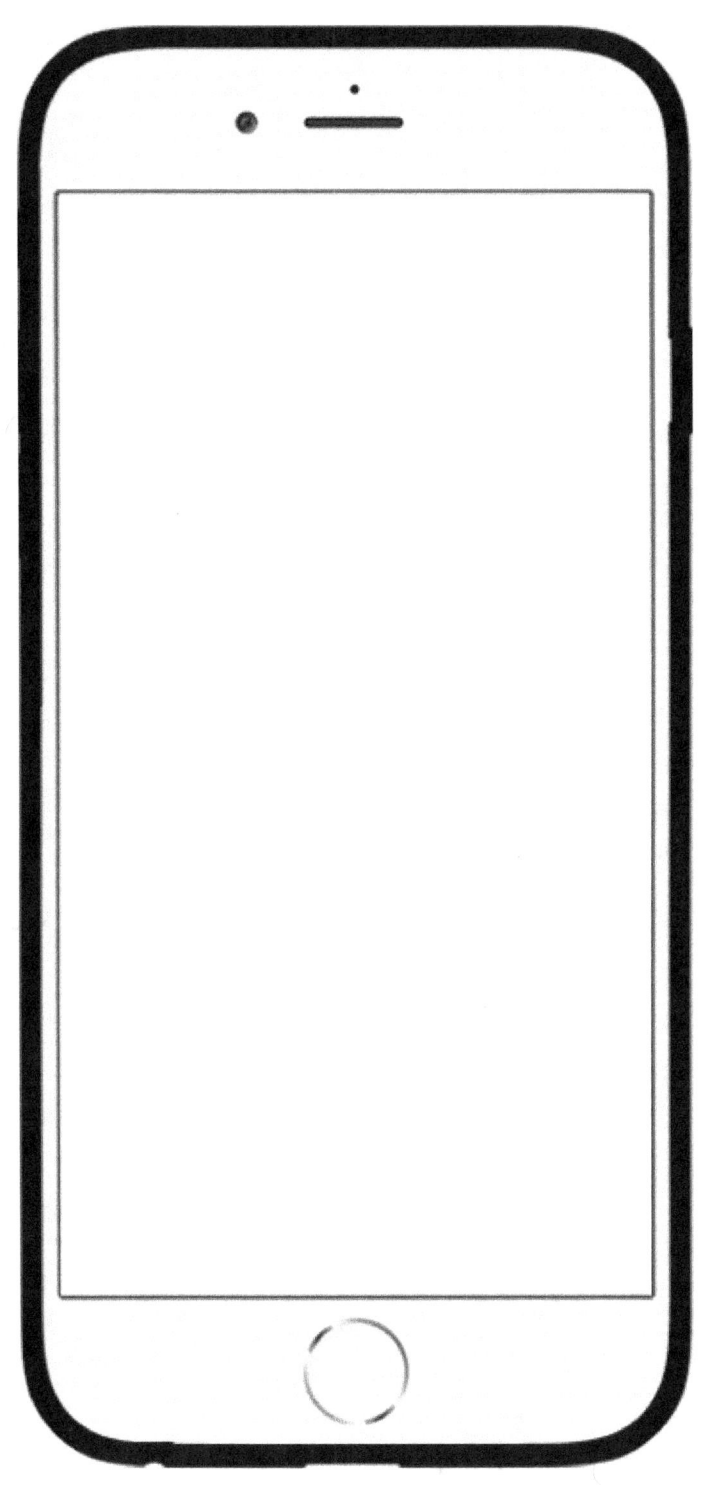

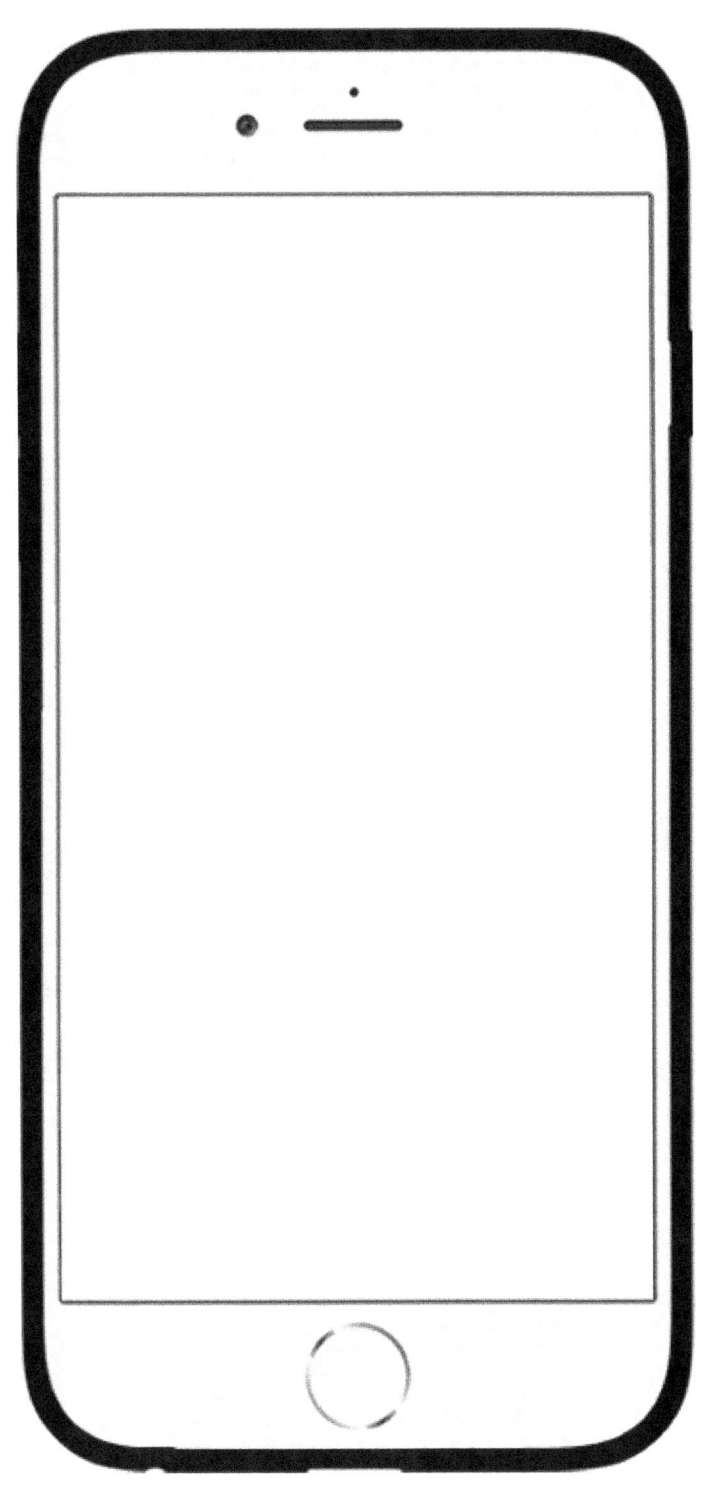

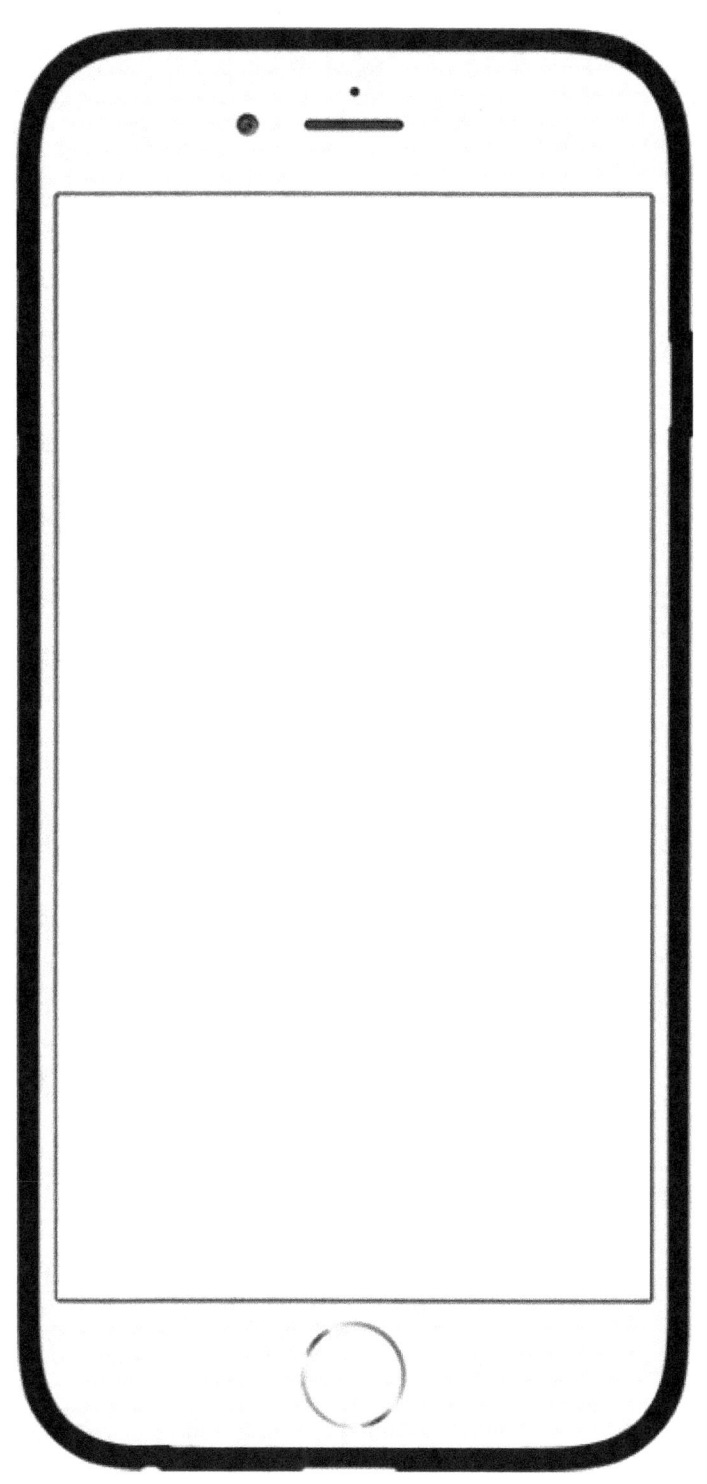

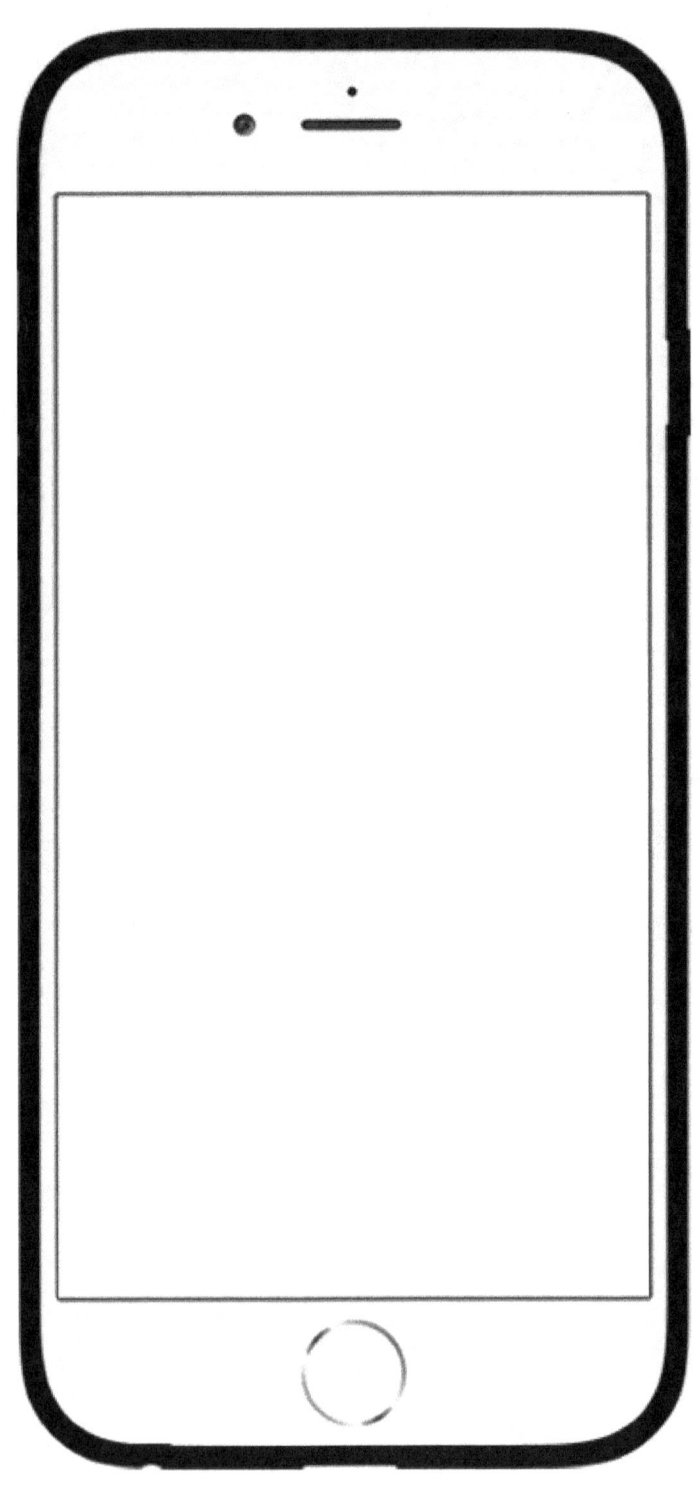

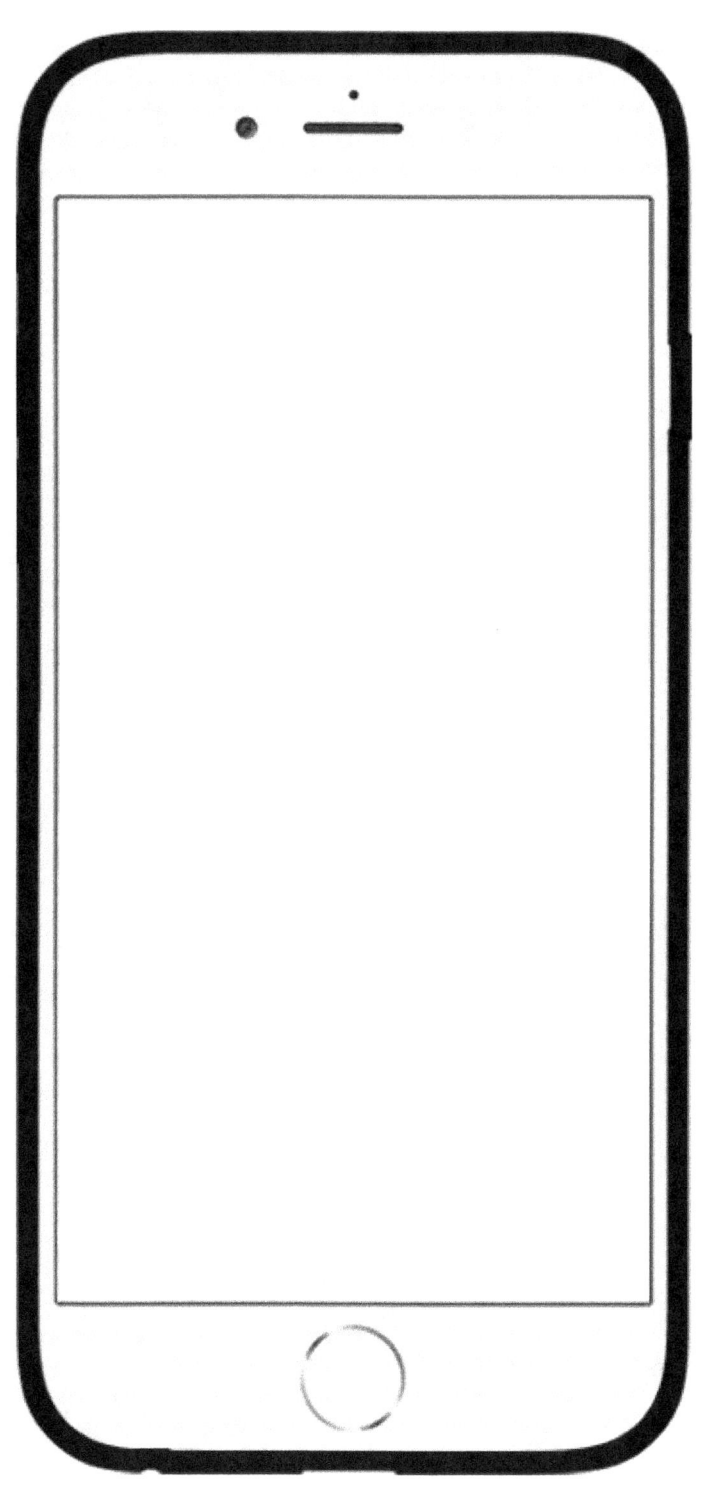

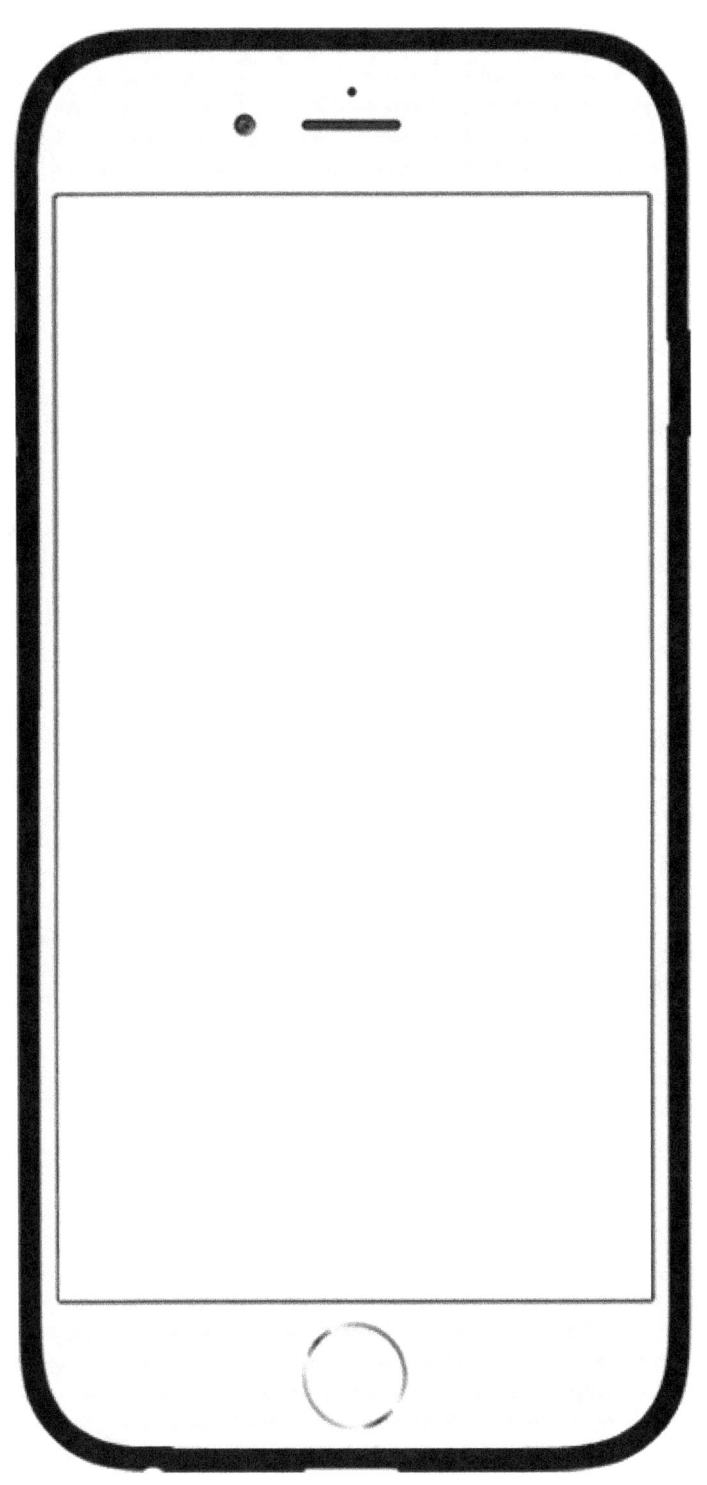

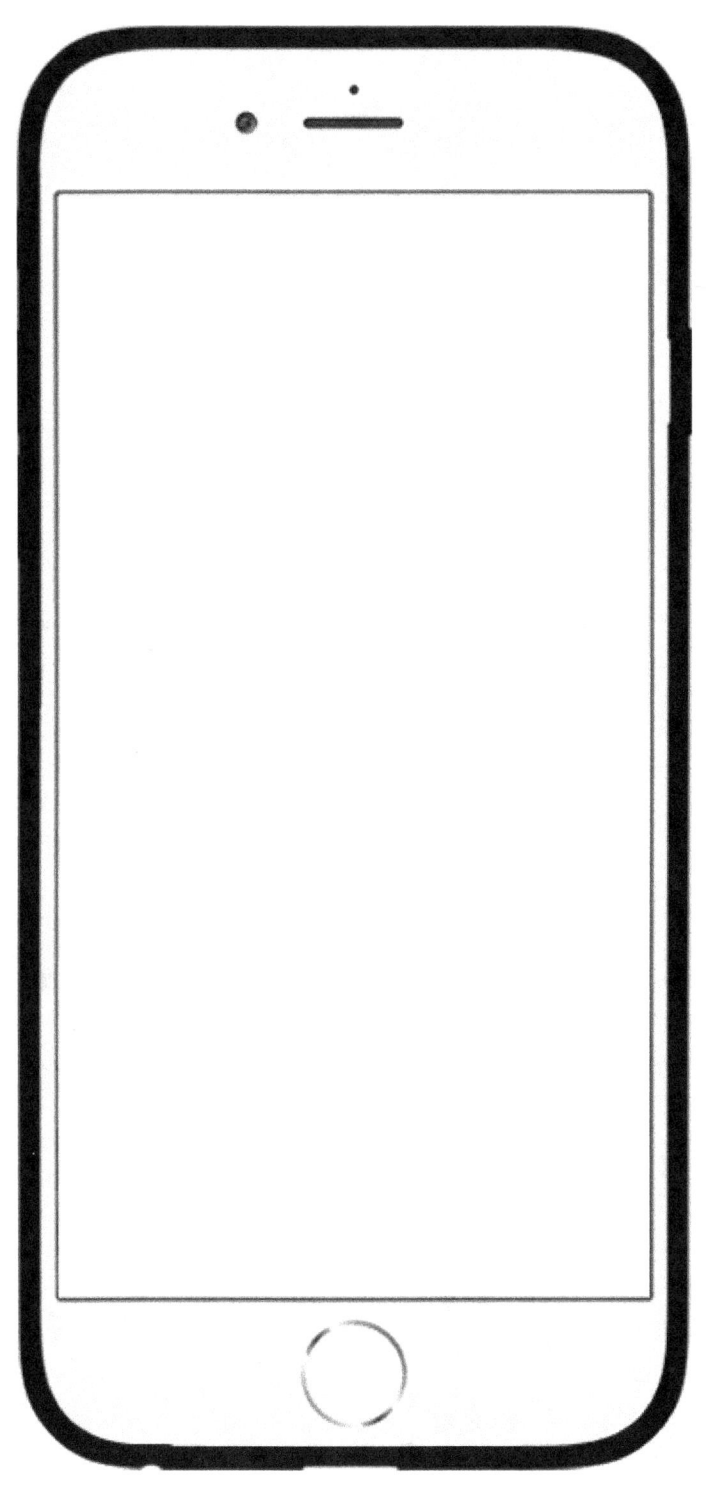

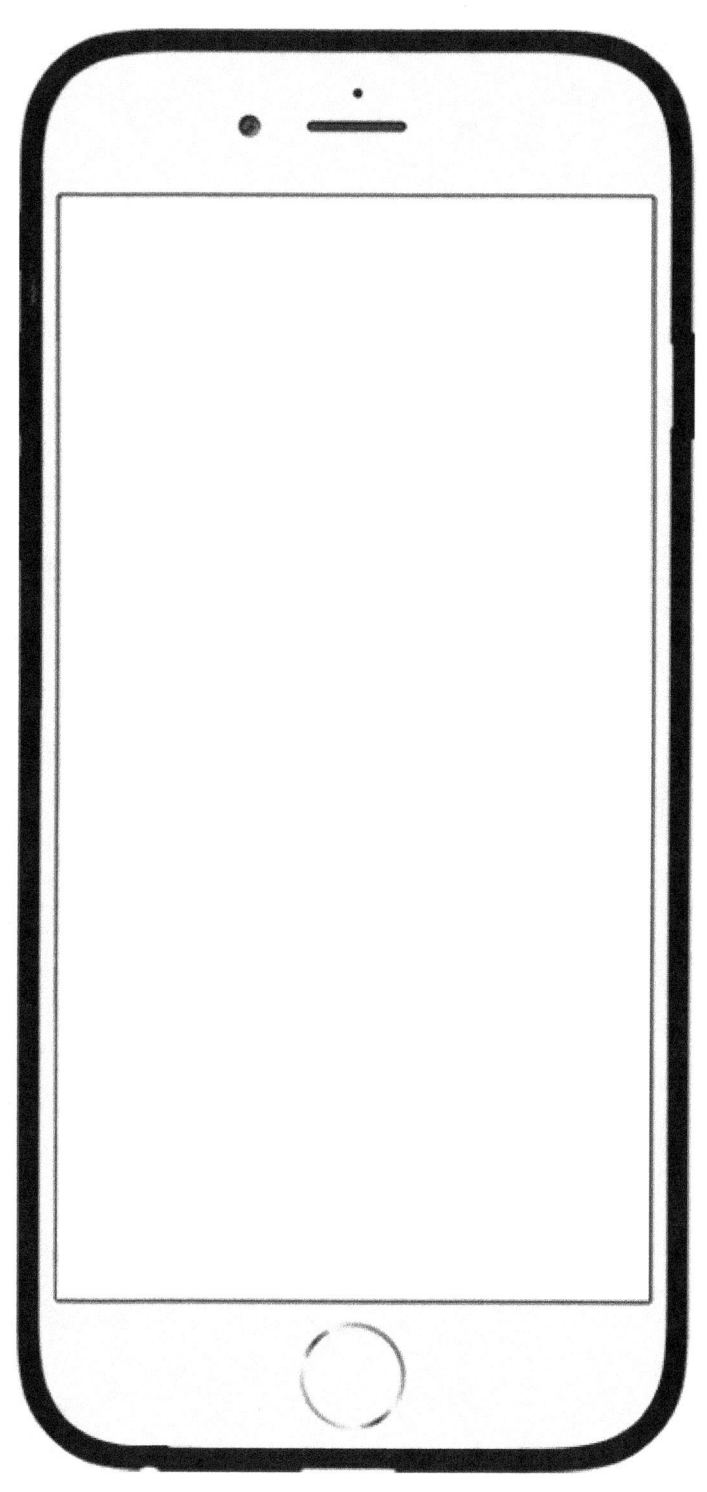

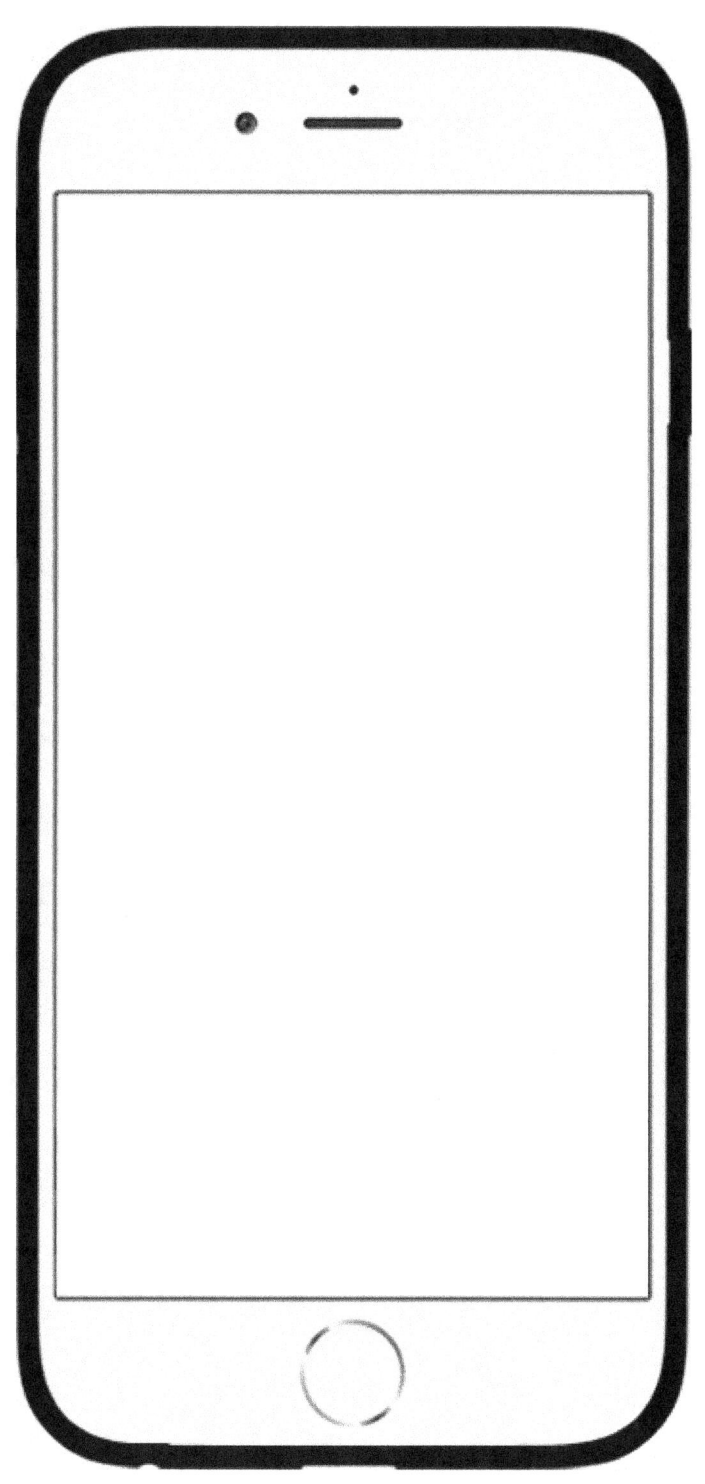

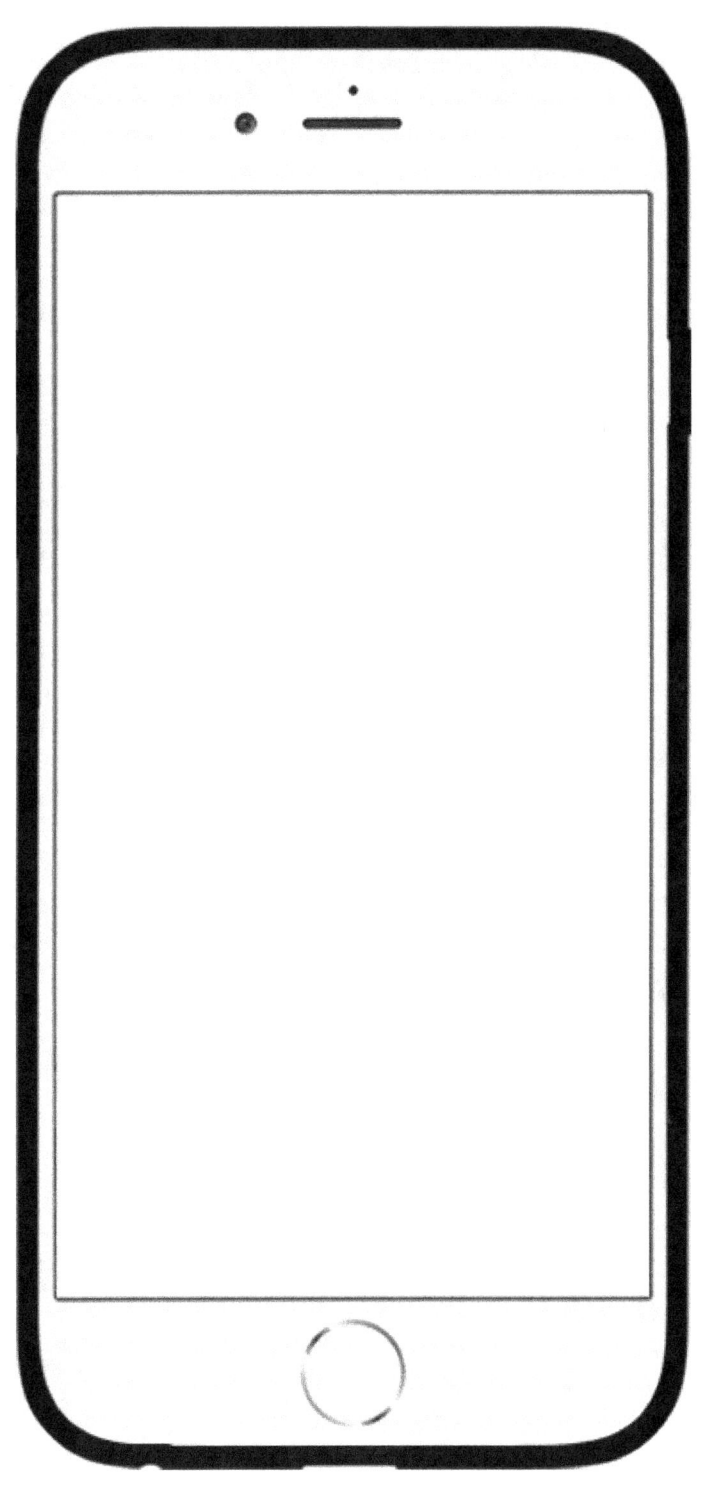

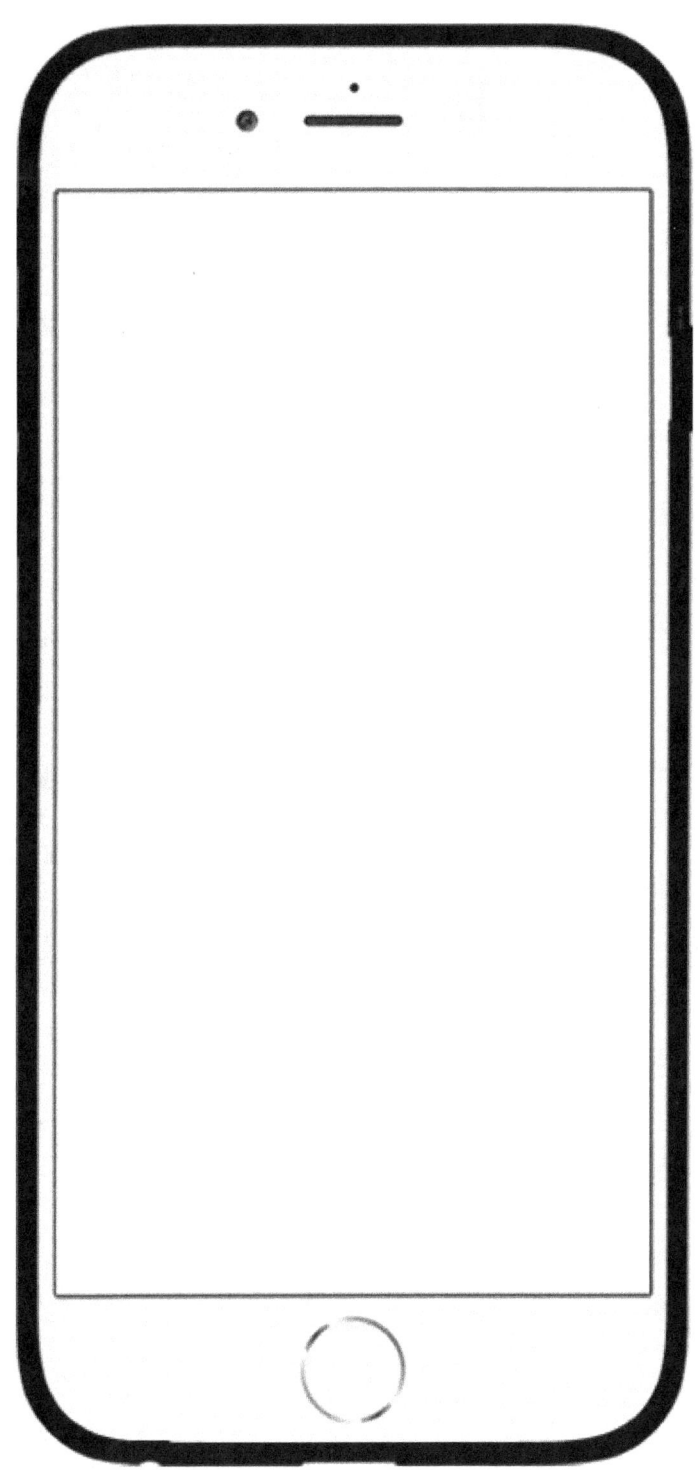

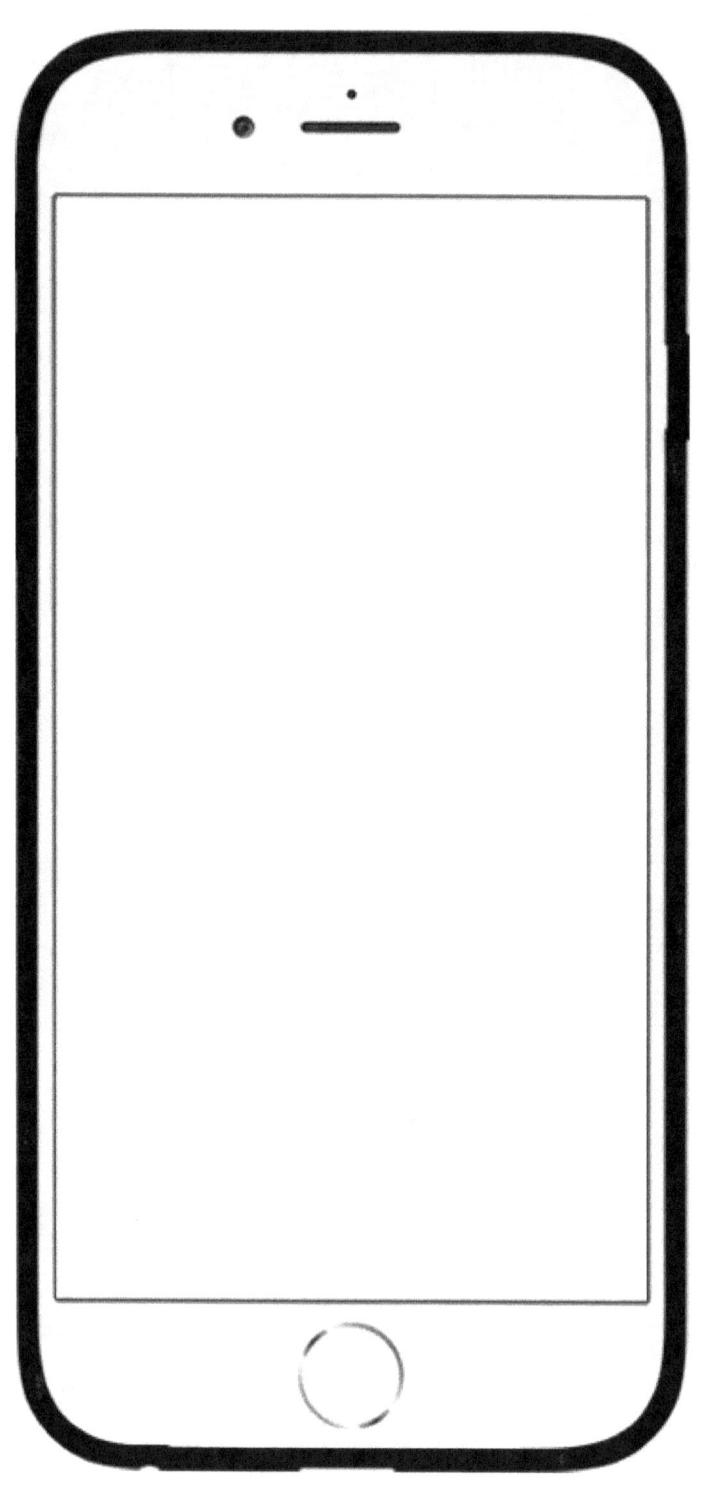

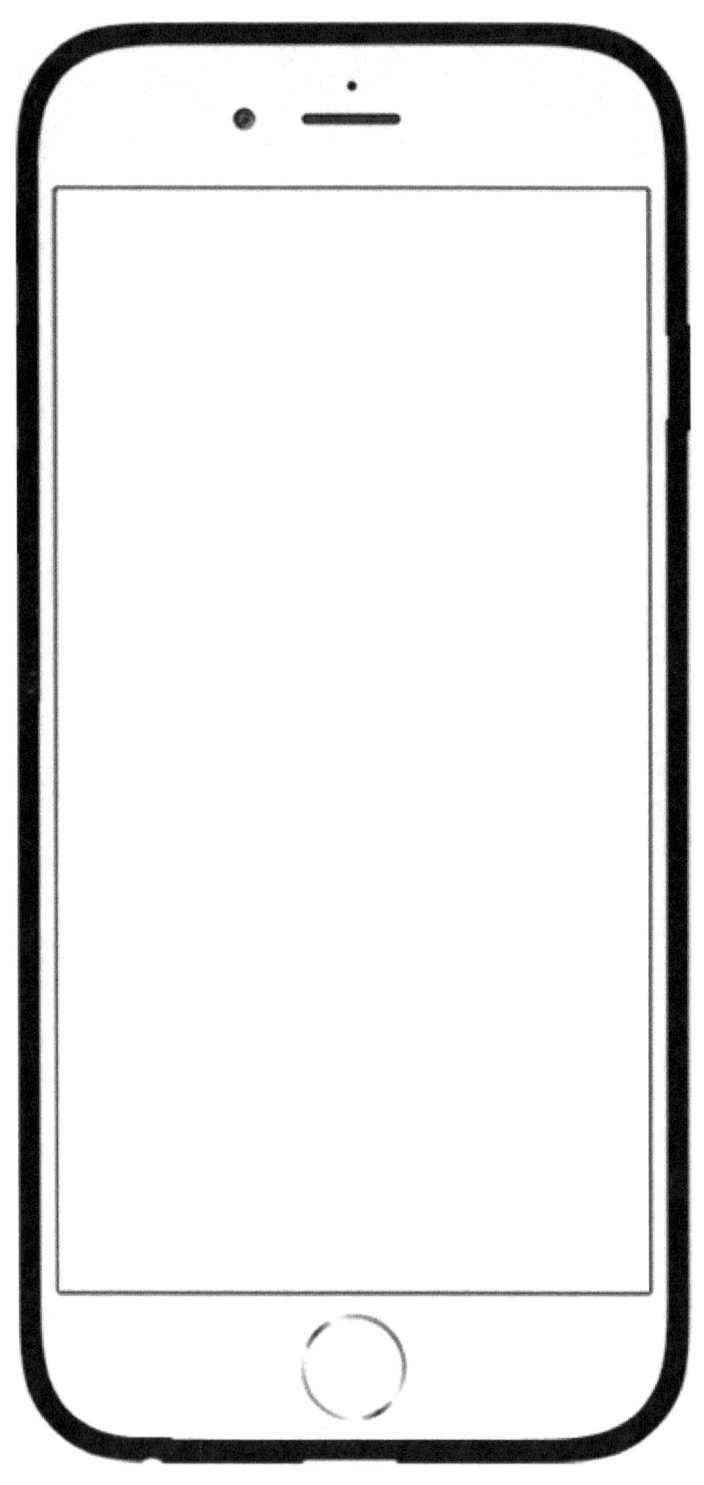

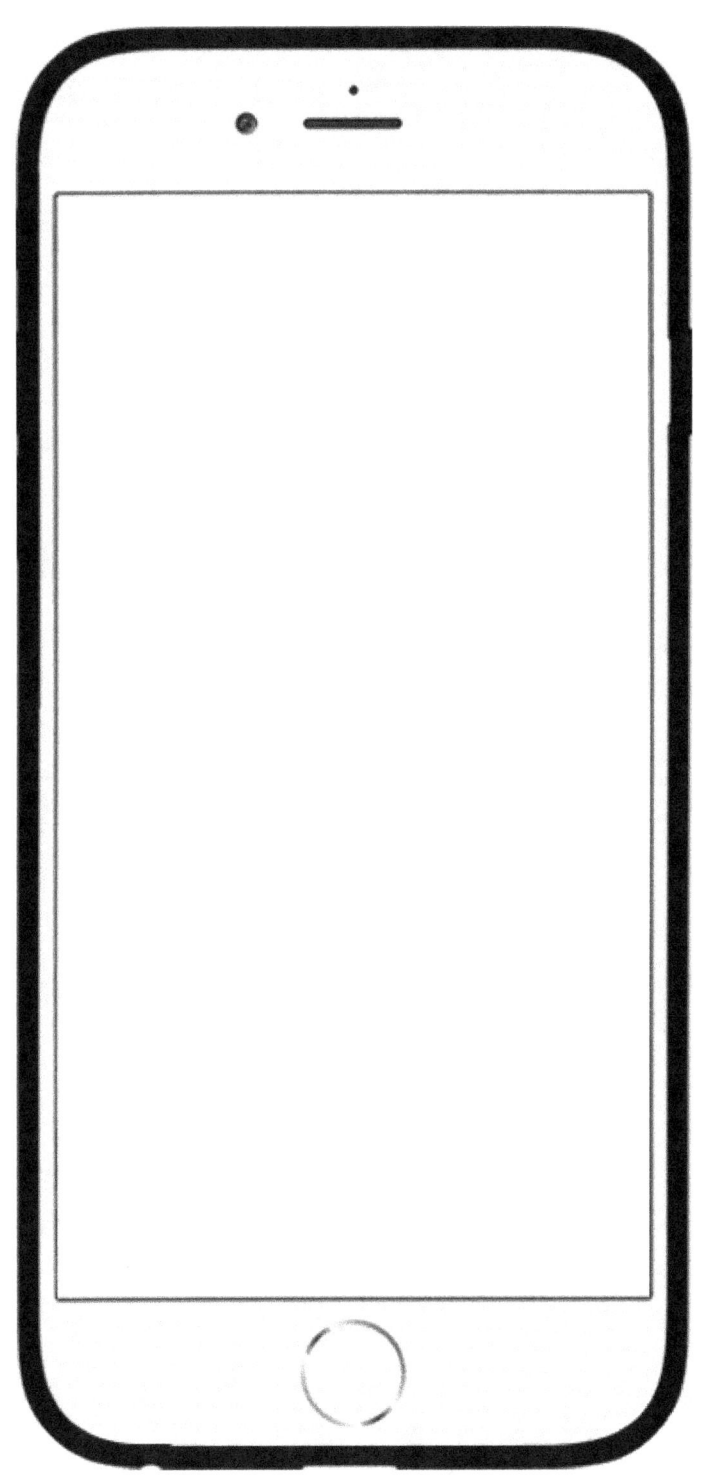

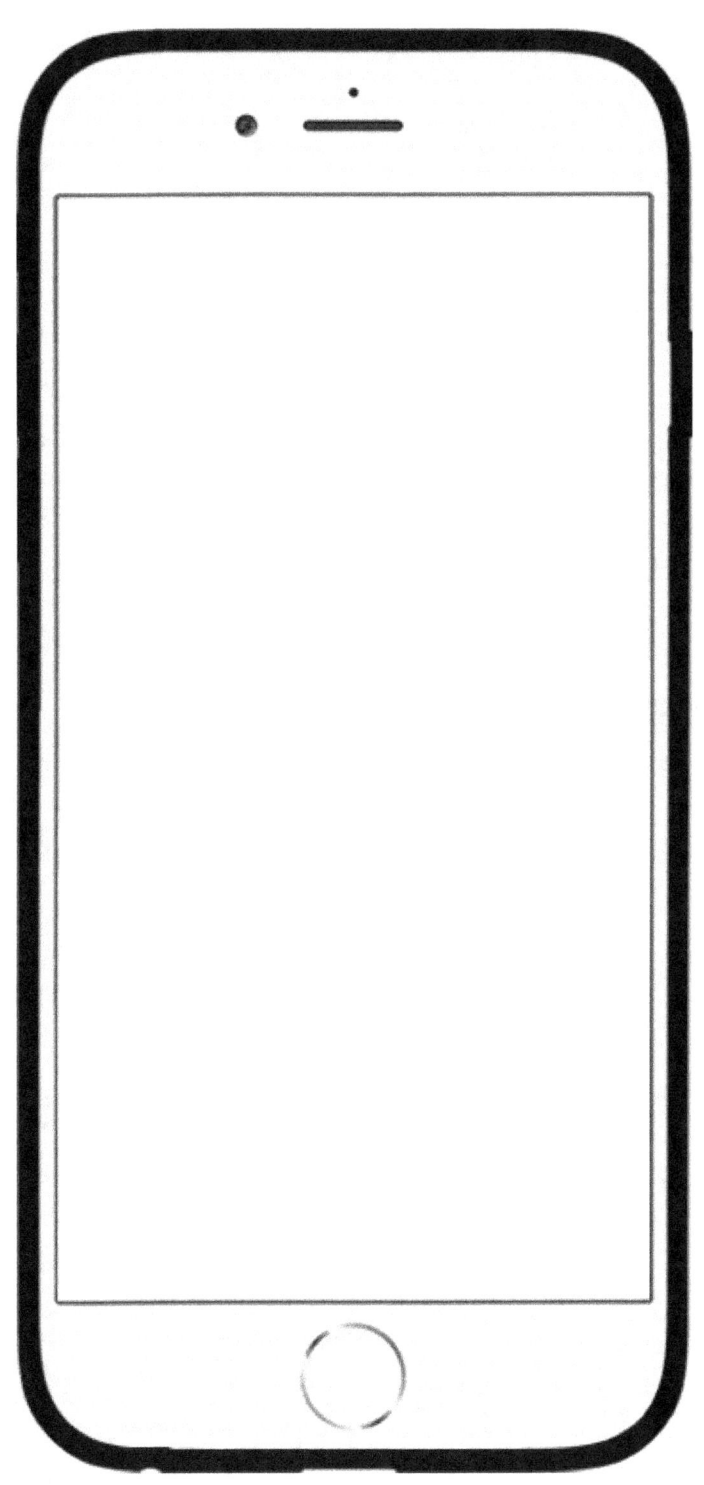

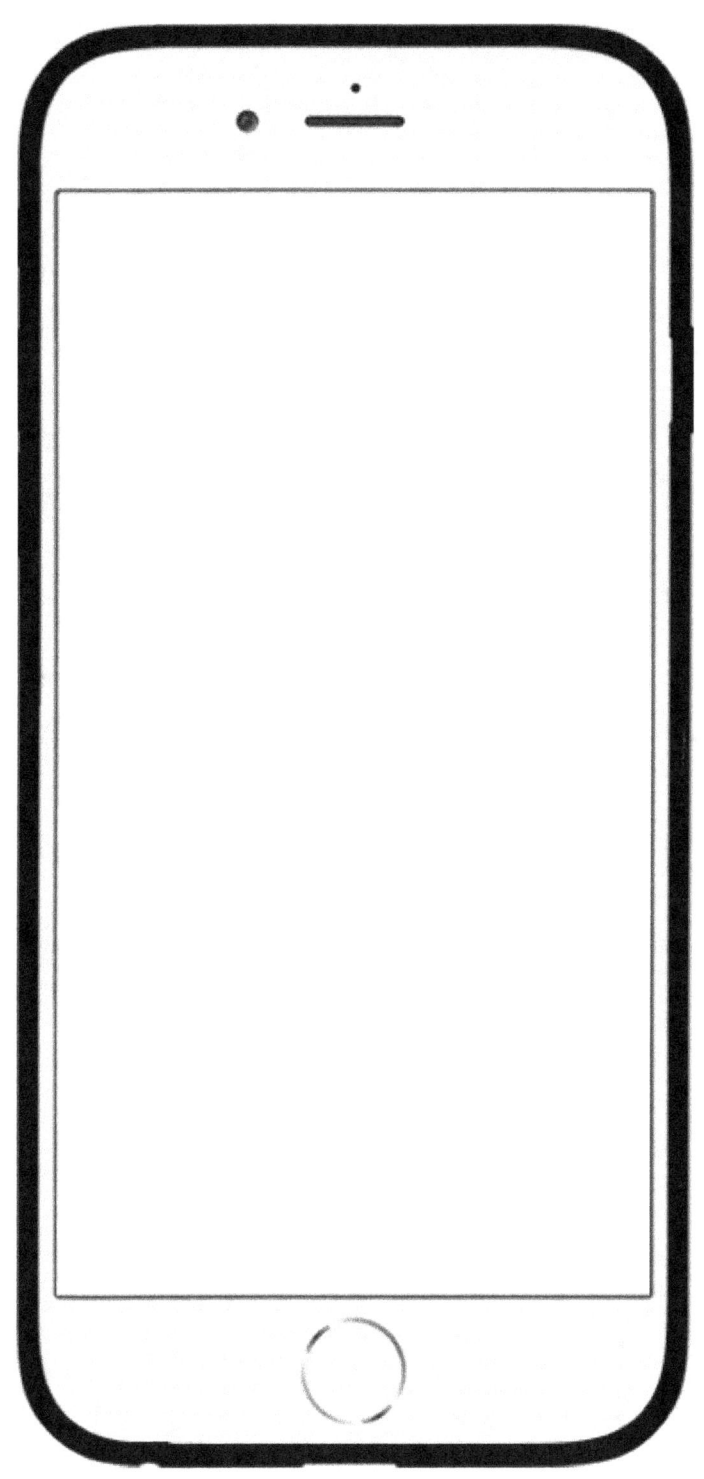

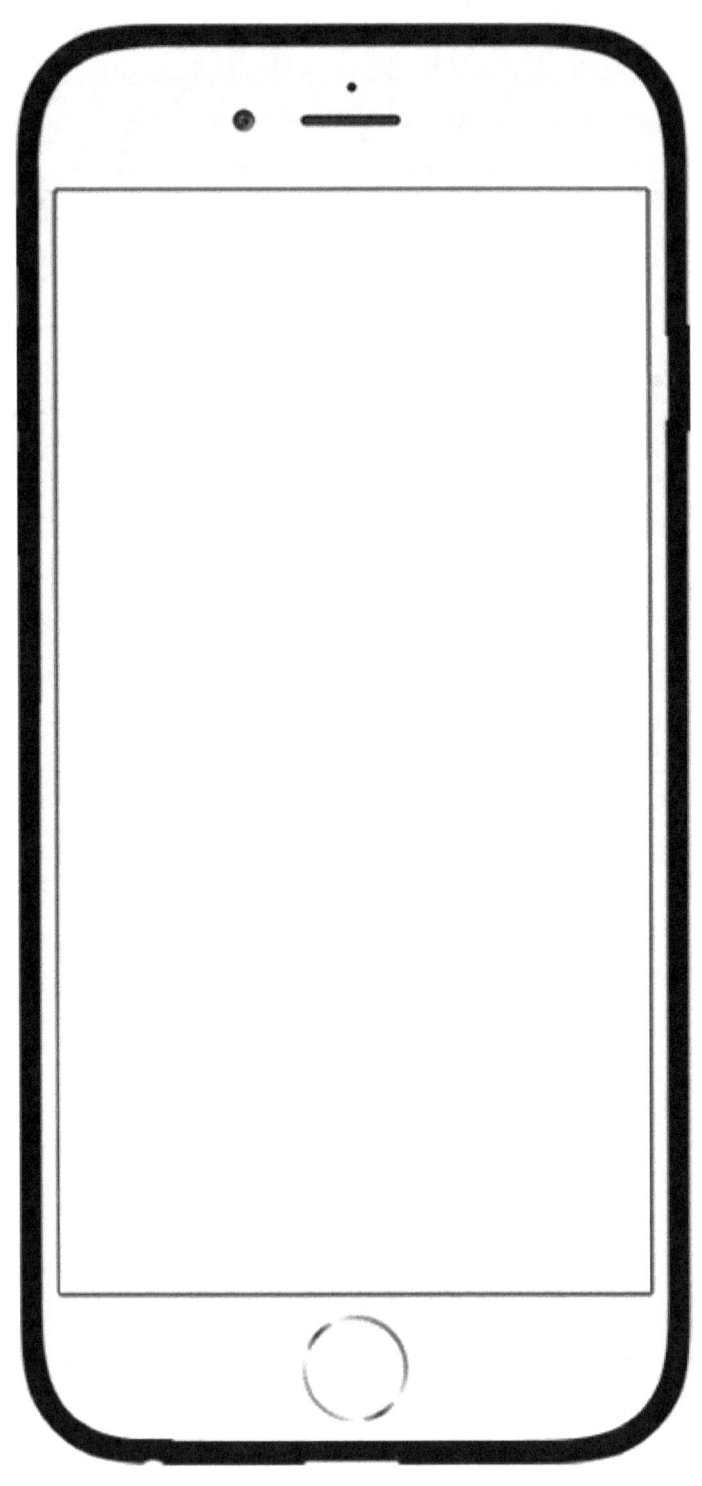

Copyright © 2018 Badapps Studio
All rights reserved.
ISBN: 1725650126
ISBN-13: 978-1725650121

www.ingramcontent.com/pod-product-compliance
Lightning Source LLC
Chambersburg PA
CBHW071409220526
45469CB00004B/1221